YOUR KNOWLEDGE HAS VALUE

- We will publish your bachelor's and master's thesis, essays and papers

- Your own eBook and book - sold worldwide in all relevant shops

- Earn money with each sale

Upload your text at www.GRIN.com
and publish for free

Bibliographic information published by the German National Library:

The German National Library lists this publication in the National Bibliography; detailed bibliographic data are available on the Internet at http://dnb.dnb.de .

This book is copyright material and must not be copied, reproduced, transferred, distributed, leased, licensed or publicly performed or used in any way except as specifically permitted in writing by the publishers, as allowed under the terms and conditions under which it was purchased or as strictly permitted by applicable copyright law. Any unauthorized distribution or use of this text may be a direct infringement of the author s and publisher s rights and those responsible may be liable in law accordingly.

Imprint:

Copyright © 2015 GRIN Verlag, Open Publishing GmbH
Print and binding: Books on Demand GmbH, Norderstedt Germany
ISBN: 9783668588264

This book at GRIN:

http://www.grin.com/en/e-book/382666/scanning-tunneling-microscope-and-atomic-force-microscopy

Suchit Sharma

Scanning Tunneling Microscope and Atomic Force Microscopy

GRIN - Your knowledge has value

Since its foundation in 1998, GRIN has specialized in publishing academic texts by students, college teachers and other academics as e-book and printed book. The website www.grin.com is an ideal platform for presenting term papers, final papers, scientific essays, dissertations and specialist books.

Visit us on the internet:

http://www.grin.com/

http://www.facebook.com/grincom

http://www.twitter.com/grin_com

Scanning Tunneling Microscope (STM) & Atomic Force Microscopy (AFM),

vacuum environment with the surface adequately cleaned of impurities and prepared as thin-film.

The success of ATM as a cost-effective imaging tool with dramatically increased ease of use has seen its proliferation to numerous labs in universities and tech companies worldwide.

Review

11th March, 2015

Suchit Sharma, Indian Institute of Technology, Dhanbad

Abstract:

Atomic Force Microscopy (AFM) is a development over Scanning Tunneling Microscopy (STM) with earlier technique only allowing conductors to be imaged. Atomic Force Microscopy has been able to image insulators also with atomic resolution by substituting tunneling current with an atomic contact force sensing arrangement, a delicate cantilever, which can image conductors and insulators alike via mechanical "touch" while running over surface atoms of the sample.

Since the sample surface contamination with foreign atoms and humidity can compromise the success of AFM, it is done in ultra-high

Article:

Imaging at atomic resolution had been an elusive goal until the introduction of scanning tunneling microscopy (STM) in 1981 by Binnig, Rohrer, Gerber and Weibel (1982). This novel approach based on the quantum mechanical concept of quantum tunneling whereby which an electron tunnels through the vacuum gap separating the biased conducting tip and conducting surface if the distance is very close i.e. atomic diameters ranges (typically 0.3-3Å).

The tunneling current being a function of separation distance, voltage difference and local density of states (LDOS) (a measure of available states per energy level in a quantum mechanical system such as an atom) fluctuates as probing tip passes over the sample surface and is converted into voltage which is mapped as imagery with the help of a computer software.

This modest instrument has provided a breakthrough in our ability to investigate and manipulate matter on atomic scale as for the first time individual surface atoms of flat surface were made visible in real time space. The invention of STM solved the most confounding problem of structure of Si (111)-(7x7) surface which was regarded as touchstone for applicability of emerging technology of STM. Takayanagi, Tanishiro, Takahashi and Takahashi (1985) complemented X-ray-crystallography with electron-scattering to STM and developed dimer-adatom-stacking fault (DAS) model for Si (111)-(7x7). Consequently, G. Binnig and H. Rohrer were awarded Nobel Prize in Physics in 1986 for their invention.

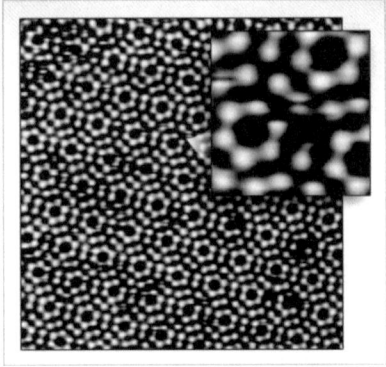

Fig 1. Si (111)-(7x7) reconstruction imaged at 1000K, Kun and Bin - The Chinese University of Hong Kong, RHK Technology, Inc.

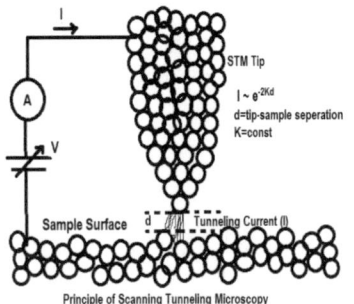

Principle of Scanning Tunneling Microscopy

Fig 2. Schematic of STR showing electron current tunneling through the potential barrier separating two electrodes (metallic or semiconductor) when distance is reduced to few atomic diameters

Quantum physics is physics will come into play predominating when distance reach microscopic scales. Classical physics is macroscopic culmination of quantum physics happening at microscopic scale. Quantum mechanics is different from classical physics in the sense that a particle is not treated as a point object as in classical physics but a quantifiable "blob" which is accordance with Heisenberg uncertainty principle can be represented by fixed area two-dimensional objects on momentum-position plane. In quantum mechanics particle nature is discovered by measurement which collapses the assigned wave Ψ(x,y,z,t) into one of the probabilities. Wave vector is change every time a measurement takes place. Hence all particles and their interaction can be considered as waves and properties associated with them. A wave continues to encompass an area where probability of finding the particle is high. When an act of measurement is done, the wave form representing the "blob" without shape or size collapses in to the particle form with observable shape and size which may go back to waveform if time is allowed. Thus act of measurement changes the wave function, collapsing it into an observable particle instantly of occurrence that fits a statistical model representing sufficiently large number of observations.

In classical physics the electron cannot penetrate a potential barrier Φ if its energy E is smaller. The quantum mechanical treatment predicts a different picture. It predicts the electron wave function will undergo exponential decaying while penetrating the barrier whilst being available on the other side. In STM a small bias voltage is applied so that due to electric field the tunneling of electron results in a tunneling current

The height of the barrier can roughly be approximated by the average workfunction of the sample and tip.

$$\Phi = \frac{(\Phi_{sample} + \Phi_{tip})}{2}$$

Electron wave function,

$$\Psi(d) = \Psi(0) e^{-2\kappa d}$$

Where, κ is inverse decay length, d is with between sample and tip

$$К = \frac{\sqrt{2m(\Phi-E)}}{\hbar}$$ here m is mass of electron, Φ is work function, E is potential applied

Since E=eV<<Φ

$$\Phi \approx 4 - 5 \ eV$$

$$К \approx \frac{\sqrt{2m\Phi}}{\hbar} \sim 1 \ \text{Å}^{-1}$$

The current is proportional to probability of electrons to tunnel through the barrier:

$$I \propto \sum_{En=EF-eV}^{EF} |\Psi n(0)|^2 e^{-2\kappa d}$$

By definition of local density of states for

$\varepsilon \to 0$

$\rho(z, E) \equiv \frac{1}{\varepsilon} \sum_{En=E-\varepsilon}^{E} |\Psi n(z)|^2$

the current can be expressed by

$I \propto V \rho_{sa}(0, E_F) e^{-2\kappa d}$

$I \approx V \cdot \rho_{sa}(0, E_F) e^{-1.025 \cdot d \cdot \sqrt{\Phi}}$

Where [d] =Å; [Φ]=eV;

Hence

$I = I_o e^{-2\kappa d}$

Also,

$I \propto V \rho_{sa}(d, E_F)$ which means that tunneling current is proportional to local energy of states at Fermi energy at distance d which is another measure for position of tip. Thus current is highly sensitive to sample tip separation

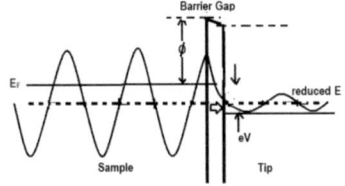

Fig 3: Electron wave overcoming a high potential barrier and register on the other side giving a tunneling current as signified by the arrow.

The tunneling current as shown depends changes with nearly one order of magnitude for 1 Å change in tip-surface separation d. This strong correlation is responsible for strong vertical sub-angstrom precision and horizontal atomic precision capability of STM. Strong distance dependence of the current ensures that the current directed to the foremost atom (nearest to the tip) is exponentially larger than the second nearest. This makes STM feasible with even blunt tips since tunneling current from atoms other than foremost atom is nearly non-existent.

The typical tunneling current is in range of pA-nA. The following graph illustrates current, I dependence on tip-sample separation d.

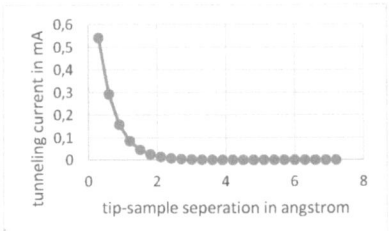

Fig 4, Tunneling-current error to tip-sample separation measurement error is high ratio which means in suitable range (the foot and upward of the slope) error in measurement of separation is very low with experimental errors in detection of tunneling current.

The error in measuring d from I i.e. $\frac{\Delta d}{\Delta I}$ as per the relation $I = I_o e^{-2.050 * d}$ is negligible at near atomic diameter distances. The error in d is in 1-5% range of an angstrom when d is in around nuclear diameter which is remarkable precision as illustrated below. Usually in an STM the tip-sample separation distance d changes by 10% while imaging which is in the range 1 Å.

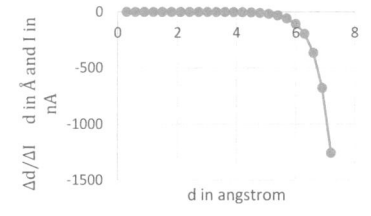

Fig 5: Error in tip-sample separation is negligible when compared to corresponding error in tunneling current measurement.

Fig 6, STM general assembly.

A close look at STM electronic modules reveals details about experimental measurement and noise in imaging.

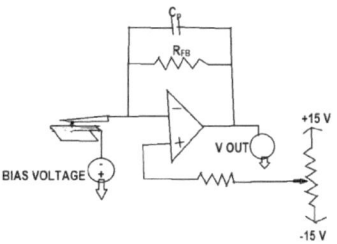

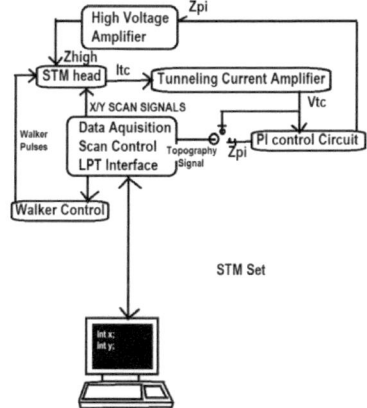

Fig 7, Op-Amp tunneling current amplifier

The typical power supply in STM assembly is +5Vdc,+/-15Vdc,+/-110Vdc and 110 Vac. Since the tunnel current high performance current-to-voltage-convertor (IVC) is an essential element of a STM. For further is in the range of pA-nA (10pA to 50nA) it is amplified via an OP-Amp Tunneling Current Amplifier. A gain of 10^9 V/A is to be created and therefore a reference a feedback type IVC is used for an ultra low input bias current and very high input impendence for high

5

voltage, wider bandwidth and lower noise is used.

The inverting feedback IVC can be built easily using commercial ultra-low input bias current and very highinput impedence operational amplifiers and a very high-ohm feedback resistor.

Ideally the inverting feedback current-to-voltage converter with a large feedback resistance above 1GΩ will provide suitable dynamic range covering the tunneling current (pA-nA) with minimized noise. A general scheme of assembly is shown. The non-inverting input of the op-amp is grounded and the voltage at the inverting input should be equal to ground. This implies

$V_{out} = -I_{IN} * R_{FB}$: R_{FB} (1 GΩ is this design, 200 pico ampere of tunneling current resulting in 200 mV output voltage).

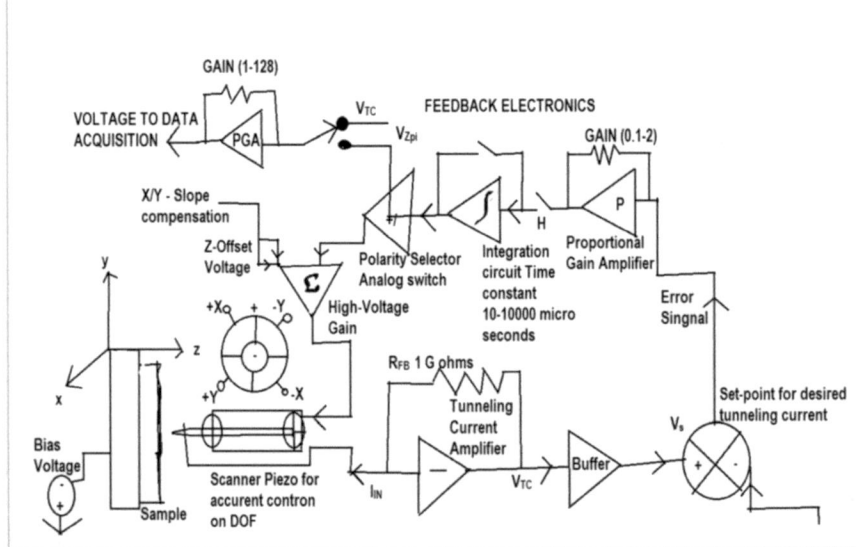

Fig.8, In STM, the feedback system is used to control imaging process. The given schematic is for fixed current STM. Initially a suitable tunneling current value is chosen and tip-sample

width is varied until the same value for tunneling current is achieved.

This is made the set-point point. A suitable frequency of movement of tip is chosen for movement in x direction which is dependent on the acquisition bandwidth. Usually with a frequency of four atoms per nanometer and a speed of 250 nanometers per second an

Fig 8: STM Assembly

acquisition bandwidth of up of 1 KHz is sufficient. As scanning begins the current is picked up by current-to-voltage converter as shown in Fig. Here it is important to understand the sources of noise in the tunneling current which although insignificant to error in z measurement is worthwhile of study.

The typical noises are Johnson noise in in the feedback resistor in the current amplifier tunneling, Johnson noise in the tunneling junction and the input noise of the operational amplifier. The Johnson noise density in a resistor is function of temperature as given:

Johnson noise, $NR = \sqrt{4k_B TR}$

k_B is Boltzmann constant, T is temperature and R is resistance. Usually the Johnson noise of all sources combined is around 0.5 pA which amounts to noise of 0.1-0.5 pm in z direction. In fixed current imagining the feedback constantly receives positive or negative error value from set-point assembly taking in buffer output from tunneling current amplifier. The feedback loop controls a cylindrical piezoelectric tube supporting the tip, expansion or contraction of which is controlled by a set of equal and opposite voltages controlling opposite quadrants when seen from circular face. Thus if a voltage difference of V is required to actuate the piezo, 2V voltage is applied across the piezo. The piezo moves the tip down or away from the sample as per the feedback from the feedback electronics.

The range of operation in z is typically in 10 % change in tip-width separation. The feedback directs the tip to move up and down to restore the tunneling current to previously set-point value. In fixed distance imaging the tip is fixed and hence registers increase or decrease in tunneling current value and consequently tip-sample separation is calculated.

If the frequency of scan in x direction is fast, a sinusoidal variation is observed else a saw teeth like corrugated signals is registered with slow scanning speed. Thus in this case contour imagery is calculated with variations in tunneling current values than actually moving on it as in case of fixed current imaging method. It is simpler in design with no feedback mechanism to control tip positioning in z direction and hence yields simple resolutions.

An important assumption taken in above analysis is linearity. The STM setup can be assumed linear except for tunneling gap. Tunneling gap shows exponential relationship between current on tip-sample width. However if tip-sample width variation is in very small range as 10% (as taken), the current shows a fairly linear behavior. For more information about assembling your own STM read Electronics and Software for Scanning Tunneling Microscope by Singh and Kumar enclosed in references.

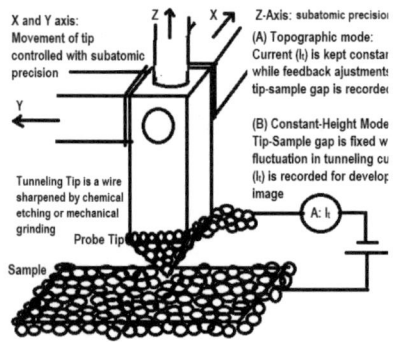

Fig9: Operation of STM with two modes (A) Topographic ($Z(I_t,X,Y)$=const.) : A suitable piezoelectric arrangement coupled to a computer keeps the tunneling current constant by varying the tip-sample gap. This fine adjustment in z axis is recorded and a three-dimensional image of sample atom (X,Y,Z) is formed.

(B) Constant-height mode: The tunneling current (I_t) fluctuates as the probe tip-sample separation is kept constant. The recorded current signal is either sinusoidal (fast-imaging) or corrugated (slow-imaging). The feedback loops lags behind spatial positioning (d,X,Y) and hence a less resolution is achieved.

The scanning tunneling microscopy has seen unprecedented success allowing atomic scale imaging of several metals and semiconductors. Several insulators surface treated with metals as gold have also been imaged yet STM suffers from this serious limitation as it works only with conductive surface.

During practice it was observed that the probing tip while in proximity with sample surface also experienced significant interatomic forces, magnitude and nature of which varied with the tip-surface gap. The attractive forces dominated initially resulting from ion-dipole attraction and van der Waals forces. However if the distance was further minimized so as the electronic orbitals overlapped, a strong repulsive ion-ion force dominated.

This was the founding principle of atomic force microscopy (AFM) which in theory can be used to image individual atoms. The basic principle of AFM is to measure forces or measure interactions between sharp probing tip and sample surface which lead to variety of other scanning probe microscopes (SPM) such as Magnetic Force microscope(MFM), the dipping force microscope (DFM), the friction force microscope(FFM) and the electrostatic force microscope(EFM). There is a tendency on unifying these measuring techniques in one spot for even greater information.

The tunneling tip of STM was replaced by a force sensing cantilever made of durable substances as Tungsten (W), crystalline silicon (Si), silicon-nitride (SiN). In ambient condition the layer of solids is under constant change, adsorbing and de-adsorbing gaseous molecules, humidity and dust. The Atomic Force Microscopy AFM is similar to STM.

A force sensor replaces the tunneling chip. The potential energy between the tip and the sample causes a Z component of the tip-sample force which is exerted equally on sample and probing tip as they are action-reaction pair but the nature and magnitude is best understood with quantum mechanics for chemical bonds. The force is because of a potential field that has a force spring constant dependent on z axis separation.

Unlike the tunneling tip which has short range contributions, electrostatic and magnetic force has both long and short range contributions, where at extremely short distances such as overlap of electronic orbitals of the probing tip atom and sample atom an increasingly repulsive force of nuclear repulsion is experienced.

In AFM, simple theoretical description and applications of forces such as ionic, magnetic, capillary, van der Waal's force of attractions, electrostatic, frictional etc. are studied and mathematical models are devised. Studies are conducted to isolate individual effect of these forces over distance but not always the individual effects are discernable as combined effect of these forces is experienced.

A simple AFM construction involves an extremely sharp probing tip attached to a cantilever type spring. In response to the force between tip and the sample, the cantilever or just lever undergoes deflection.

Images are taken by scanning the sample relative to the probing tip and digitizing the deflection of the lever or the z-movement of the piezo as a function of lateral position x,y. The typical spring constants are in between 0.001 to 100N/m and motions from microns to 0.1 Å are detected by the deflection sensor. Forces range from 10^{-11} to 10^{-6} N. For comparison a covalent bond is at 1 Å with a force of 10^{-9} N. Hence non-destructive imaging of covalent bonds, in case polymers and bio-materials is possible.

Two force regimes exist for AFM, a non-contact regime and a contact regime. In non-contact regime the tip-sample separation is at 10-100 nm so forces as van der Waals, electrostatic, magnetic and capillary forces are imaged. At smaller distances as 1 Å, strong ionic repulsive forces dominate and permit imaging of the atomic boundary. In addition frictional forces and elastic or plastic deformation can be detected under appropriate conditions.

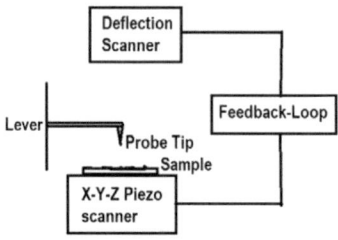

There are two modes of operation of AFM, static mode or the dc mode and dynamic mode or ac mode. In static mode, the cantilever is allowed to deflect till equilibrium is achieved and force is calculated by Hooke's law.

Force $F = c_b z_t$

where c_b is spring constant and z is displacement. A rectangular beam with a constant cross section is

$c_b = 3EI/l^3$ where E is the young's modulus, I in moment of inertia and l is length. A beam with width b and thickness $I = bd^3/12$

With dimension of $1*10*100$ μm³ of a rectangular Si-cantilever ($E = 1.69*10^{11} N/m^2$) a spring constant of $c_B = 0.42$ N/m is arrived. In static mode force in range of $10^{-10} - 10^{-6}$ N can be measured. In equiforce mode the deflection in pin is kept by moving the sample up or down.

A different mode called variable deflection mode, requires the tip-sample distance be fixed and variation of deflection made in the lever to digitized for determining deflection force. It has high scanning speed compared to previous method but less precise. In dynamic mode, the lever is allowed to oscillate close to its resonance frequency

Most AFM in universities, public and private research centers use ambient conditions with reduced resolution for simplicity and avoiding added complexity in maintaining and use of AFM apparatus in ultra-high vacuum conditions. Though latter gives extraordinarily sharp resolution without distortions in actual surface view due to foreign atoms and humidity.

An AFM is designed to measure the force in between the sample and the cantilever. Often techniques in electron tunneling, optical interferometry, laser beam deflection and capacitance method for detecting change of force. Tunneling aparatus is sensitive to contaminants and tunneling gap become comparable to force detection gap between the tunneling tip and the force side of the lever. Thus other techniques such as optical interferometry, laser beam deflection and capacitance methods are introduced.

Unlike tunneling apparatus, these methods detect force over a gap on nm to centimeters, hence deflection is stable. In optical techniques, light pressure is used for detecting interaction which is in the order of 10^{-12} N. In capacitance methods, electron-static forces are too weak in order of 10^{-11} N and hence there is almost no interaction between the sample and cantilever for most applications.

Miniaturization of optical methods is limited by the wavelength of light, which gives insight into the limits of force microscopy.

In dynamic mode or ac mode the lever is oscillating close to its resonance frequency. A distance dependent force F(z) affects the resonance curve. The equation of motion of rectangular beam is given by:

$$\frac{\partial \Psi}{\partial x} + \mu/EI \frac{\partial \Psi^2}{\partial x^2} = F(x,z)$$

Where μ is the mass/length (m/l)

Now $\Psi = Y(x).T(t)$

The time part is $T^2+\omega^2 T=0$

The nth Eigen-frequency is given by

$$F_n = \frac{(\kappa_n l)^2}{2.\pi l^2}\sqrt{\frac{EI}{\mu}} = \frac{(\kappa_n l)^2}{2\pi\sqrt{3}}\sqrt{\frac{c_B}{m}}$$

Where $f_n = 2\pi\omega_n$ and κ_n depends on the space dependent part and therefore also depends on the force acting on the tip. For the above mentioned Si-cantilever the resonance frequency of its first Eigen mode is found to be 138 kHz.

The force gradient $F' = \frac{\partial F}{\partial z}$ influence the resonance frequency. The effective spring constant is then

$c_{effective} = c_B - F'$.

A repulsive force stabilizes the spring and increases the resonance frequency. An attractive force does the reverse , lowers the resonance frequency

$$f_1 \approx 0.32\sqrt{\frac{c_{eff}}{m}} = 0.32\sqrt{\frac{c_B - F'}{m}}$$

In ac mode either the amplitude is kept constant while detecting the slope or Amplitude is measured keeping the tip-sample separation constant, i.e. FM detection.

The fixed gradient of resonance measuring or the fixed amplitude mode requires a feedback loop. Also the scanning speeds are slow because amplitude if regained with sample base x-y-z scanner piezo. A simple and faster design but more difficult to interpret is fixed separation method that requires no feedback loop and change in amplitude is recorded as height contour. The typical forces measured lie in the force range F(z) = const. z^{-n} and values are between 10^{-5} to 10 N/m. Z is around 10 mm with forces between 10^{-13} to 10^{-7} N.

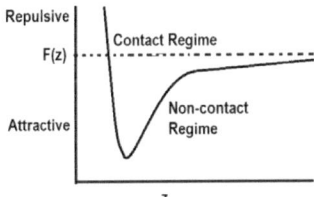

The imaging signals are described by the root-mean-square (rms) deviation of the mean value and indicated by the prefix δ,

$\delta V \equiv \sqrt{<(V-<V>)^2>}$ where V is the voltage output from current amplifier.

Quantum mechanical treatment of chemical bonds has been done and models have been devised. An H_2^+ ion is a model for the covalent bond. The Morse potential describes a chemical bond with bonding energy E_{bond}, equilibrium distance s, and a decay length k for the H_2^+ ion chemical bond.

$$V_{morse} = -E_{bond}(2e^{-\kappa(z-\sigma)} - e^{-2\kappa(z-\sigma)})$$

The above equation describes a chemical bond with bonding energy E_{bond}, equilibrium distance σ, and a decay length κ. However above cannot describe an important nature of covalent bonds called anisotropy i.e. the chemical bonds are highly direction or angular.

Another potential describing effect of van der Waals forces of attraction and repulsive forces of contact is Lennard-Jones potential.

$V_{lennard\text{-}Jones} = -E_{bond}\ (2*\frac{z^6}{\sigma^6} - \frac{z^{12}}{\sigma^{12}})$, here proportionality to σ^{-6} represents attraction due to van der Waals forces and $-\sigma^{-12}$ term represents repulsion due to contact forces.

The Si-subatomic structure can be explained using Stillinger-Weber potential. It was worked with ab-initio calculations and accounts for the neighbor and the next-neighbor interaction which is necessary for unstable arrangements as those of diamond like in Si.

The nearest-neighbor contributions in Stillinger-Weber potential is

$V_B(r) = E_{bond}A[B(r/\sigma')^{-p} - (r/\sigma')^{-q}]*e^{1/(r/\sigma'-a)}$

for r<aσ' , else $V_{nn}(r)=0$

The next-nearest-neighbour contribution is

$V_{nn}(r_i,r_j,r_k) = E_{bond}[h(r_{ij},r_{ik},\theta_{jik})$ + $h(r_{ji},r_{jk},\theta_{ijk}) + h(r_{ik},r_{kj},\theta_{ijk})]$

with

$h(r_{ik},r_{kj},\theta_{ijk}) =$

$\lambda e^{\lambda(1/\frac{r_{ij}}{\rho'-a} - \rho'-a/r_{ik})}*(\cos(\theta_{jik}+\frac{1}{3})^2$

for $r_{ij,ik}$ <aσ', else 0.

Ab-initio calculations for modelling can be done now a days with advanced computers.

Instead of measuring force versus distance , force gradient vs distance is measured in ac mode and numerical integration of force gradient w.r.t z over z gives the force distance curve.

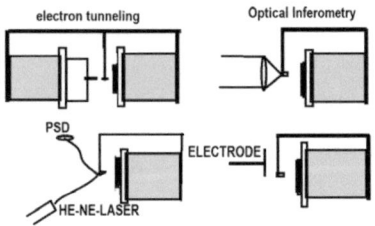

When talking of the deflection sensors it is important to keep in mind that the first sensor was based on electron tunneling. However the conditions of electronic tunneling gap may be contaminated by films and dirt and interaction with tunneling tip and the force sensing cantilever which can often become comparable to tunneling tip-surface interaction make the technique sometimes too sensitive. A less sensitive optical interferometry relies on detecting light reflected from the probing tip back. It measures the light pressure. Another technique employs lasers instead of ordinary light. Yet another technique uses capacitance to measure tip-sample separation forces. In all the above techniques tip-sample interaction is stronger than tip-measuring assembly set-up.

The interferometry techniques is advantage up to the scales of wavelength of a laser and calibrations of piezos. It measures the photocurrent produced by the striking of photons of light and

$I = | E_O^2 + E_R^2 | = I_O + I_R + \sqrt{I_O I_R} \cos(\Phi)$

Where Φ is phase difference between object and reference frame waves. The object wave is $\Phi = \Phi_o + 4\pi \Delta z/\lambda$ where Δz is deflection of lever and λ is wavelength of light source. used. The variation of Δz is recorded by a piezoelectric positioning device and relative variations of the intensity of the photocurrent is

$\frac{\Delta I}{I} = \frac{2\sqrt{I_O I_R}}{I_O + I_R} * 4\Pi \Delta z/\lambda$

Which can be reduced to

$\Delta I/I = 4\Pi \Delta z/\lambda \approx \frac{\Delta z}{\lambda}$

In case of beam deflection optical methods, the cantilever is polished on the rear side so that a reflected beam is incident on a photo-sensitive detector (PSD), two or four-segmented photodiode. The difference between intensity from upper and lower halves of the diode divided by the total intensity gives a direct measure of the deflection of the lever as:

$\frac{\Delta I}{I} = \frac{I_A - I_B}{I_B + I_B} = \frac{6\Delta z}{l} * 1/\delta$

Where l is length of the lever and δ is the beam divergence of the reflected beam. In case of capacitance methods small change in distance causes a relative variation of 10^{-5} in capacitance which is detected by change of voltage which is comparable to optical methods.

$\frac{\Delta U}{U} = \frac{\Delta z}{z}$

Capacitance of a plate type capacitor is given by $C = \varepsilon_r \varepsilon_0 A/z$ where A is area of plate and z is the distance between plates.

When choosing cantilevers it is important fro them to extremely sensitive to forces and hence their spring constant should be in-between 0.01-100 N/m. In order to minimize the effect of acoustic waves and vibrations leading to resonance, the resonance frequency is kept high at about 10-100KHz.

Now resonance frequency is proportional to cB/m. Therefore, the mass of the cantilever has to be minimized hence the dimensions need to be miniaturized. Typical dimensions of thickness, width and length are 1 x 10x 100 m³ which gives for a S-lever a resonance frequency of 138kHz and a spring constant of 0.42N/m.

Scanning electron micrograph of a micro-machined silicon cantilever with an integrated tip pointing in the [001] crystal direction. In this type, the tip is etching free so that the sample area adjacent to the tip is visible in an optical microscope. Length, 120 mm; width, 30 mm; thickness, 2.8 mm; c$_B$ 515 N/m; f 300 kHz. Photo courtesy of Olympus Optical Co. Ltd, Hachioji, Tokyo, Japan.

Conclusion:

Scanning tunneling microscopy and atomic force microscopy are revolutionary inventions which have greatly enhanced out view of the subatomic world and research on untapped properties of materials in designing next generation of materials with new properties. Information technology will hold the key in how we decipher the information we receive at atomic scales, deciphering it and appropriately manipulating in it advanced materials will be the advance of materials science and engineering.

References:

- Baym, G., 1969, Lectures on Quantum Mechanics (Benjamin, New York)
- E Meyer, institute of Physics, University of Base/, Klingelbergstrasse 824056 Base/, Switzerland
- Franz J. Giessibl, Experimentalphysik VI, Electronic Correlations and Magnetism, Institute of Physics, Augsburg University, D-86135 Augsburg, Germany
- Franz J. Giessibl, "Advances in atomic force microscopy." Reviews of Modern Physics, 2003
- Hug, H. J., M. A. Lantz, A. Abdurixit, P. J. A. van Schendel, R. Hoffmann, P. Kappenberger, and A. Baratoff, 2001, "Technical Comment: Subatomic features in atomic force microscopy images," Science
- Leon van Dommelen, Fundamental Quantum Mechanics for Engineers, 5/5/07 Version 3.1 beta 3
- Partin, J., 1995, "Atomic resolution of an insulator by

- noncontact AFM," Presentation at the 12th International Conference on Scanning Tunneling Microscopy 1995 (Snowmass, Colorado, USA).
- Pethica, J. B., and R. Egdell, 2001, "The insulator uncovered,"Nature (London)
- Phil Russell, Appalachian State University, AFM Probe Manufacturing, courtesy Oliver Krause, NanoWorld Services GmbH
- Principles of Atomic Force Microscopy (AFM)by Arantxa Vilalta-Clemente Aristotle University, Thessaloniki, Greece and Kathrin Gloystein Aristotle University, Thessaloniki, Greece based on lectrures by Prof. Nikos Frangis Aristotle University, Thessaloniki, Greece
- Robert W. Carpick, Miquel Salmeron, "Scratching the Surface: Fundamental Investigations of Tribology with Atomic Force Microscopy". Chemical Reviews 1997
- Robert A. Wilson and Heather A. Bullen, Department of Chemistry,Northern Kentucky University, Highland Heights, KY 41099 Introduction to Scanning Probe Microscopy
- Garcia Ricardo, Armin Knoll, Elisa Reido, "Advanced Scanning Probe lithography." Nature Nanotechnology 2014
- Stanley Cohen, MD AFM methodology review in American Society for Investigative Pathology June 21, 2016
- Yong Chen, Jiye Cai, Meili Liu, Gucheng Zeng, Qian Feng and Zhengwei Chen research on Double-probe, Double and Triple Tip effects during Atomic Force Scanning PMC 2010 May 4
- For general reading Springer Open and Wikipedia, the online resource at https://www.springeropen.com/ and https://www.wikpedia.org/ respectively was referred

YOUR KNOWLEDGE HAS VALUE

- We will publish your bachelor's and master's thesis, essays and papers

- Your own eBook and book -
  sold worldwide in all relevant shops

- Earn money with each sale

Upload your text at www.GRIN.com
and publish for free